PREFACE TO THE FIRST EDITION

Effective monitoring of the ecological factors is one of the prerequisite of all the environmental studies and practices. Methodological literature on the examination of physical and chemical parameters of water is so disseminated that a worker in this field is often confused in search and choice of correct method. There is hardly any wide-ranging literature to cover the problem entirely. In the present effort, I have attempted to present the analytical methods as well as the general introduction, importance and hazards for a wide array of physical and chemical factors pertaining to water.

The subject matter of this book comprises of 25 chapters. The book starts with the chapter one describing different sampling methods and thereafter chapter 2 to 25 describes about analysis of different physiochemical factors and their environmental significance.

Although a large number of environmental factors are tried to be covered, but still cannot be regarded as the definitive. Methods presented in this book are not appealed to be my original, but are taken from numerous authentic sources to which I am thankful. Little modifications in the methods and calculations are made wherever felt necessary in order to make them simple to follow and apply. Not all the methods given here represent to the latest of available ones but the selection is made on the basis of practicability and ease at no or minimum cost of precision of result. Chemical reactions and principles involved are also included wherever possible to describe every matter completely. References to the authors are not made in the text but are included as bibliography at the end.

30/06/2015   Vineeta Girdoniya

Contents

Chapter – 1   **Sampling of Water**

A- Collection of water samples from water resources for physiochemical analysis.

1- Sample containers:

Containers shall be made of chemically resistant glass like Pyrex. Before using them for the collection of water samples they shall be cleaned with chromic acid to remove all extraneous surface dirt. Sample bottles shall be provided with either glass stoppers or plastic caps.

2- Volume of sample:

For most of the physical and chemical analysis, 2.5 liter portion is adequate. Immediate field analysis is required for certain constituents and physical characteristics like temperature, pH, DO, CO_2, residual chlorine, etc., to assure dependable results, since changes in composition of the samples may occur during the transit to the laboratory. For DO determination, a small sample (300 ml) may be collected separately in a BOD bottle and fixed on site with MnSO4 and alkali azide solution.

3- Time interval in between the sampling and the analysis:

A potable water specimen may ordinarily be held for a much longer period than a raw waste water sample. Following maximum limits are suggested by standard methods as reasonable for physical and chemical analysis:

Clean waters (Non polluted)   -   72 hours.

Slightly polluted water   -   48 hours.

Polluted water   -   12 hours.

4- Frequency of sampling:

For reasonable accurate estimate of the quality of water, it will be collected monthly and analyzed.

5- Point of sampling:

Ponds/ Lakes:

Where the water in a stream is well mixed so as to approach uniformity, a sample taken at any point in the cross section is satisfactory.

River:

Point shall be away from the shore, in case of the water works, point shall necessarily be from the upstream area with respect to the intake well.

Open Wells:

Sampling shall be made rather rapidly under the surface of water preferably 30 cm below the surface.

Municipal supplies/ Tap Water:

Sample shall be collected directly from tap.

B- Sampling for Bacteriological Assessment:

1. Sample Containers:

Samples should be collected in clean, sterilized, neutral glass bottles of 250 ml capacity. These sampling bottles should be provided with a ground glass stopper. The stopper and neck of the bottle shall be protected by strip of paper during sterilization. The bottle shall be sterilized in a hot air oven at 160° C for one hour or in a autoclave at pressure of at least 1.054 kg/cm^2 (15 lb /in^2) for 20 minutes. The sampling bottle should not be opened except at the time of sampling.

2. De-chlorination:

If the water to be sampled is likely to be contains chlorine, 0.1 ml of sodium thio-sulphate solution (3 percent) shall be added to bottle before sterilization. When samples of chlorinated water are taken, it is desirable to determine the content of residual chlorine in the water at the sampling point.

3. Time Interval Between Sampling and Analysis:

Bacteriological testing of water shall be commenced as soon as possible after collection. Where it is not possible, the sample shall be kept in ice or in the refrigerator till they are taken up for analysis. All such iced or chilled samples shall be taken up for analysis within 48 hours after collection.

4. Procedure for Sampling:

Sampling in River, Stream, Lake, Pond etc.:

Samples should not be collected from a place far away from the point of draw off. The bottle with the stopper and cover removed. And shall be held by the bottom and plunged neck downwards below the surface of the water. The bottle shall then be turned until the neck points slightly upwards with the mouth directed along the current.

In the case of lake, pond, swimming pool, reservoir, the current shall be artificially created by pushing the bottle horizontally forward in a direction away from the hand.

Well Water:

String shall be attached to the neck of the bottle and shall be fully wrapped in paper and then sterilized. The paper cover shall be removed just before taking sample, avoiding any contact to sample water. Another clean long string shall be tied to the sterilized string and the bottle lowered into the water and allowed to fill up. The bottle shall then be raised the stopper with cover replaced and the string removed.

Tap Water:

Tab shall be opened and allow water to waste for 2 minutes, and then turn off. It is then sterilized by heating with an ignited piece of cotton soaked in methylated spirit till it becomes unbearably hot. The tap then shall be cooled by allowing water flows from this, and the bottle is filled avoiding splashing. Samples should not be collected from leaky taps. Avoid touching sample water with finger or any other thing, inside the neck of bottle or the corresponding outer face of the ground glass stopper or to lay stopper down so that this face comes in contact with any surface, because this may result in contamination of sample. The bottle should not be filled completely. A small air space shall be allowed below the stopper so as to keep little oxygen for the respiration of bacteria.

5. **Sampling Points:**

Since bacteriological analysis controls the safety of water to the users. Location of sampling points will be more as far as possible, especially at pumping stations, treatment plants, reservoirs and booster pumping stations, as well as, various places in the distribution system.

6. **Frequency of Sampling:**

Table- 1: For untreated waters entering the distribution systems-

S. No.	Population Served	Maximum interval between Successive Samples
1	20,000	1 Month
2	20,001- 50,000	2 weeks
3	50,000-100000	4 weeks
4	More than 100000	1 day

Table- 2: For water collected from distribution system, whether treated or not-

S. No.	Population served	Maximum Interval Between Successive Samplings	Minimum Number of Samples to be Taken from Entire Distribution System
1	20,000	1 Month	One sample per 5000 of population per month
2	20,000-50,000	2 Week	--
3	50,000-100,000	4 Days	--
4	More than 100,000	1 Day	One sample per 10,000 of population

2 Sampling for Biological Examination:

General Considerations:

The microscopic organisms, other than bacteria in water include a large variety of algae, fungi, yeast, protozoa, rotifers, crustacean, etc. Many of which affects the quality of the water for drinking and industrial uses. The organisms which occur free-floating in water are collectively known as planktons, while those which occur attached to a substratum are called benthos. The examination of the nature and number of the organisms, present in the sample is of use in understanding the nature of the pollution, cause of undesirable tastes and odors, slime growth etc. Some algae are also known to produce substances that are toxic to animals.

Samples should be examined immediately after collection, when the organisms are alive. If this is not possible, the samples have to be preserved in ice or in the refrigerator for a few days taking care not to allow them to freeze. If the examination is to be made later, samples shall be preserved by adding a 3 to 5 % solution of formalin.

Procedure:

The sample of the natural plankton producing water shall be collected in clean, neutral glass bottles of two liter capacity, fitted with ground glass stopper. The bottle will not be filled completely and a small air space shall be left below the stopper. A concentrated sample of the plankton organisms and other particulate suspended matter in the water shall be collected with the aid of plankton net. The net shall be conical in shape, of suitable size, with a circular mouth and made of bolting silk cloth with more than 6000 meshes per square centimeter. The net shall be hauled through the water in an oblique or horizontal direction for a certain distance, lifted from the water, allowed to drain and the organisms in the net washed down into container by splashing water on the outer surface of the net. The catch shall then be made up to a known volume with the original water. Nets provided with closing devices shall be used for collecting samples of the plankton from different depths. Samples of sludge and bottom deposits shall be collected by means of a dredge or a scoop sampler.

Frequency of Sampling:

Sampling frequency is based on the circumstances prevailing but generally one sample per season will suffice.

Information to be supplied with the sample:

(i)   Name and address of person of institution sending samples for examination.
(ii)   Date and time of collection and dispatch.
(iii)   Purpose of examination.
(iv)   Source of water.
(v)   Information regarding sampling points and the procedure adopted for sampling.
(vi)   Temperature of water.
(vii)   Possibility of water getting affected due to recent rainfall or under particular circumstances.
(viii)   Information regarding source:
       (a) Well- Depth of well, and of water surface from ground level. Whether covered or not. Nature, material, and type of well construction.

(b) River- Nature of the flow, information about floods, whether level of water is above or below normal? Presence of bathing Ghats, boat jetty, burial ground or sewer outfall in the near proximity?

(c) Lakes/ Ponds- Catchment area whether conserved or not? Nature and extent of weed growth.

(d) Service reservoir- Open or covered. Frequency and method of cleaning.

(ix)   No. of hydrants and sewers on the distribution system.

(x)   Hours of pumping and supply.

(xi)   Population served.

(xii) Any other particular.

Introduction

Natural waters remain partly or wholly in motion due to action of various factors like wind, in or out flow, and silt etc. These movements affect the transfer of matter and energy in the system. In polluted waters currents have important role in the course of self-purification.

I-   Pilot Tube Method

Requirements:

> Pilot tube, a simple L- shaped glass tube open at both the ends. One end of this tube is shorter than the other. Longer axis of the tube may be progressed in centimeters or its fractions.

Method

> Dip the pilot tube in flowing water vertically, with the shorter end submerged and facing the flow. Water enters the open end and rises a distance h above water level outside in the vertical limb of the tube. Measure the distance h.

Calculation

$$V = 0.977\sqrt{2gh}$$

> Where, v= velocity (m/sec); 0.977 = an average constant; g = acceleration of gravity (9.8 m/sec); h = height of water column (in fractions of a meter).

> To take capillary into account, dip the upper end of the tube into water and measure the rise of water due to capillarity. Subtract this value from h.

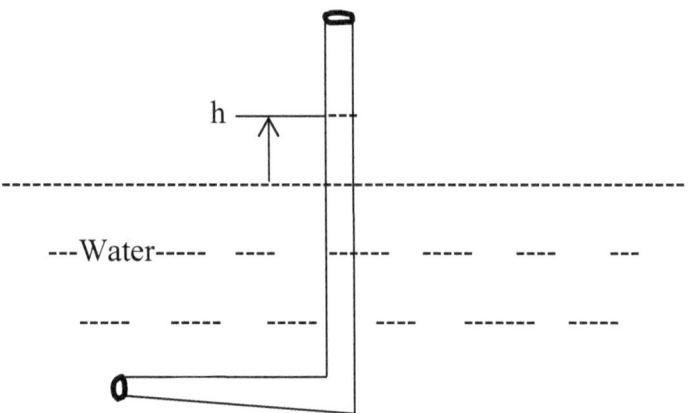

Fig.1 Pilot tube

II- Float Method

Requirements

   a- Float (cork or wood); b- Magnetic compass; c- stop watch; d- Measuring tape etc.

Method

Release a float on the surface of the water and note the direction of flow of water with the help of a magnetic compass. Fix two poles at a known distance (d) in the direction of flow. Now release the float at one pole and note the time (t) taken in reaching to the other pole using a stop watch. Repeat the exercise for several times and calculate the mean of time.

Calculation

$$V= \frac{d}{1.2\,t}$$

Where, v = velocity (m/sec); d = distance between the two poles (m); 1. = a constant; and t = time (second).

III- Flow meter method

Modern flow meters are very suitable and precise instruments for measurement of flow of water. These can be lowered at any depth of water to measure the flow at that level. Flow meters are also attached with plankton net to know the actual amount of water passed through the net.

Introduction

By color in the water means those hues inherent within the water itself which result from colloidal substances and materials in solutions. In natural waters, color may occur due to the presence natural vegetation like phytoplankton, weeds, or organic extracts like tannins, humic acids, fulvic acids, suspended matters, metallic ions, decomposition products of lignin's and industrial effluents, etc., that can be removed by coagulation.

I-   **Forel- Ule color scale method-**   This method gives a qualitative recognition of the color of the sample.

Requirements:

(i)   Solution A- Add 0.5 g of $CuSO_4$. $5H_2O$ to 5 ml strong NH_4OH and dilute to 100 ml with distilled water.

(ii)   Solution B- Add 0.5 g of $K_2CrO_4.5H_2O$ to 5 ml of strong NH_4OH and dilute to 100 ml with distilled water.

(iii)   Solution C- Add 0.5 g of $CoSO_4.7H_2O$ to 5 ml of strong NH_4OH and dilute to 100 ml with distilled water.

Empirical Forel- Ule color scale is prepared by mixing different proportions of above solutions (solution A, B and C) as shown in the table.

Method

Centrifuge the sample to remove the suspended matter. Fill the sample in a clean tube and compare its color with the 22 mixtures of the Forel- Ule color scale. Find the matching color and note the color of sample from this scale.

Table-3: Forel- Ule Color Scale.

S. No.	Solution	I	II	III	IV	V	VI	VII	VIII	IX	X
i	A	100	98	95	91	86	80	73	65	56	46
ii	B	0	2	5	9	14	20	27	35	44	54
iii	C	0	0	0	0	0	0	0	0	0	0
	COLOR	Blue		Greenish blue		Bluish green			Green		

S. No.	Solution	XI	XII	XIII	XIV	XV	XVI	XVII	XVIII	XIX	XX	XXI	XXII
i	A	35	35	35	35	35	35	35	35	35	35	35	35
Ii	B	65	60	55	50	45	40	35	30	25	20	15	10
iii	C	0	5	10	15	20	25	30	35	40	45	50	55
	COLOR	Greenish yellow					Yellow				Brown		

II- Color determination using Photoelectric Colorimeter:

Requirements-

Potassium chloroplatinate – Dissolve 1.246 g potassium chloroplatinate K_2PtCl_6 (equivalent to 500 mg metallic platinum) and 1 g crystalline $CoCl_2.\ 6H_2O$ in 100 ml concentrated HCl and dilute to 1 liter with distilled water. Stock solution = 500 units.

Procedure:

Dilute the standard solution so as to match color with sample and find out the units by comparing with color discs. Apparent color can be measured by using suitable filters in the photoelectric colorimeter.

III- Platinum cobalt method- This method gives a quantitative measure of the shade.

Requirements:

(i) Nessler tube (50 ml)

(ii) Color standards- Dissolve 1.245 g of potassium chloroplatinate and 1.0 g of crystalline cobaltous chloride in a little distilled water; add 100 ml of concentrated hydrochloric acid and dilute to 1 liter with distilled water. This solution has a color value of 500 color units. Prepare standards having color value of 5, 10, 15, 20, 25, 30, 35, 40, 45, 50, 55, 60, 65, and 70, by diluting 0.5, 1.0, 1.5, 2.0, 2.5, 3.0, 3.5, 4.0, 4.5, 5.0, 5.5, 6.0, 6.5, and 7.0 ml of above solution with distilled water to 50 ml in standard Nessler tubes. Indicate on each tube the color value, and protect them from evaporation and contamination.

Method

Centrifuge the sample at high speed to remove suspended matter. Fill standard Nessler tube with sample to same level as that of standard (50 ml) Compare the color of the sample with that of various standard tubes held vertically above a white surface and find one standard which has same shade as that of sample and read its color value. If the sample shows a color greater than 70 units, it should be diluted with distilled water. In such cases, original color is calculated as:

Color (unit) = $\quad$ Estimated color $\times$ Dilution water

Chapter-4   **Turbidity**

Introduction

Absorption coefficient of a liquid or it is the expression of the optical property of a sample which causes light to be scattered and absorbed rather than transmitted in straight line through the sample. This is because of the presence of some factors like planktons, clay, and silt, finely divided inorganic and organic matter.

Suspension of particles in water interfering with the passage light is called turbidity. Turbidity is caused by wide variety of suspended matter, which range in size from colloidal to coarse. Turbid water is undesirable from aesthetic point of view in drinking water supplies. Turbidity affects light scattering, absorption properties and aesthetic appearance in a water body. Increase in the intensity of scattered light results in higher values of turbidity.

The source of turbidity can either be physical (e.g., fine sediment particles), biological (e.g. algal blooms), chemical (e.g., precipitation of ion oxide).The nutrients like nitrogen and phosphorus enriches water bodies which ultimately causes increase in algal blooms, making the water body more turbid. During rainy seasons water bodies becomes more turbid due to due to dissolution of sediments in it. But when water becomes stagnant, its turbidity decrease because of settled down of dissolved matter. The settle-able solids include inorganic and organic non dissolving solids, which settle to the bottom of the ponds causing siltation of water resources (Verma, 1978).It is expressed as Jackson Candle Turbidity unit (JCT) or Net Turbidity Unit (NTU).

Measurement

Turbidity was determined with help of turbidity meter (Nephlo–Turbidity meter-132(Make-Systronic)). It is expressed as Net Turbidity Unit (NTU).

Procedure

Samples were shacked thoroughly. Placed rigid until air bubbles disappear and then, poured into turbidity meter tube. Readings for turbidity noted directly from instruments scale. More turbid samples were diluted with distilled water and then the readings were taken.

Chapter- 5　　　　　　**Transparency (Light Penetration)**

Introduction

Solar radiation is the major source of light energy in natural waters in that it gets transformed into potential energy by biochemical reactions. Thus governing the primary productivity of aquatic system it indirectly controls the whole biological set-up inside the body of water. Depth to which the light attenuation is 1% of what received on surface is called compensation level where photosynthesis of the system is considered to be equal to respiration. Zone above compensation level is the trophogenic zone where photosynthesis is in excess of respiration, while the zone below this is the tropholytic zone where respiration is in excess of photosynthesis.

Transparency of water is inversely proportional to the turbidity, which in turn is directly proportional to the amount of suspended organic and inorganic matter.

I- Secchi disc method

Requirements-

Secchi Disc: It is a metallic disc of 20 cm diameter with four quadrants on upper surface painted black and white alternately. It has a hook in the center of upper part, to which a graduated cord is tied. On lower surface it has a centrally placed weight which facilitates the sinking of the disc in proper position. This device is designed by A Secchi in 1865.

Method

Lower the Secchi disc in water and note the depth (cm) at which it disappears. Now raise the disc slowly and note the depth at which it reappears.

Calculation:

$$\text{Sec chi disc transparency (cm)} = \frac{A + B}{2}$$

Where, A = Depth at which Secchi disc disappears (cm); and B = Depth at which Secchi disc reappears (cm).

$$\text{Euphotic limit (cm)} = SDT \times 2.5$$

$$\text{Vertical attenuation coefficient} = \frac{1.9}{SDT}$$

Where, SDT = Secchi disc transparency (cm).

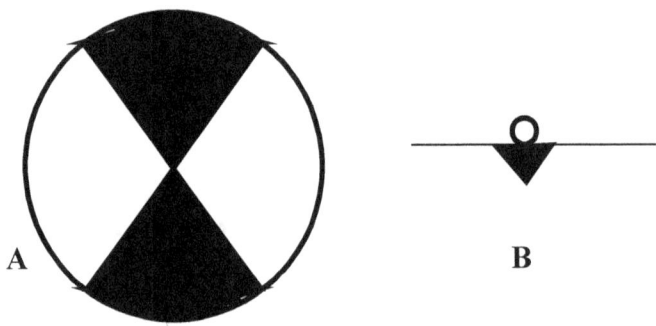

Figure-2: Secchi Disc. A- Top view; B- Side view.

Introduction

Imposing solar radiation and the atmospheric temperature bring about interesting spatial and temporal thermal changes in natural waters which manifest in setting up of convection currents and thermal stratification. Temperature is a measure of intensity of heat in terms of a standardized unit. It is an extremely important, the easiest to measure, and most extensively studied universal ecological factor. It greatly influences the vital activities like metabolism, behavior, reproduction, and development. Organisms tolerant to a wide range of thermal fluctuations are called eurythermal and those to a narrow range are called stenothermal.

Temperature fluctuations and extremes are more produced in the air than in the water medium. In nature, environmental factors do not operate independently of one another, and temperature is one such factor which largely influences, and is influenced by, other factors like moisture and wind. Diurnal and annual ranges of atmospheric temperature are more important in any ecological investigation rather than

Water temperature is very important factor, because it influences the biota in a water body by affecting activities such as behavior, respiration and metabolism. Temperature plays a very important role in wetland dynamism affecting the various parameters such as alkalinity, salinity, dissolved oxygen etc. All organisms possess well defined limits of temperature tolerance. Temperature is also important because of its influence on water chemistry, for example, warm water holds less oxygen then cool water. Temperature also played important roles in increase of metabolism and physiological reactions of organisms.

1- Surface water temperature

Requirements: A mercury thermometer of 0 to 50° C range and with 0.2° C least count.

Method:

Take surface water in a plastic container and record its temperature immediately by dipping the thermometer for about one minute.

2- Sub-surface temperature

A- Using thermos- flask sampler

Requirements

A sampler made up of thermos flask and having an inbuilt thermometer.

Method

Lower the sampler in water to the desired depth and close it. Draw the sampler out of water and read the temperature.

B- Using reversing thermometer

Requirements

A reversing thermometer consists of a large reservoir of mercury connected with a distal bulb by means of a fine capillary. Just above the reservoir this capillary is constricted and has a side branch at this point, above which it makes a loop and then becomes straight to join the bulb. The volume of mercury above the constriction depends upon the temperature and when the thermometer is reversed at 180° C with the help of a messenger, the mercury thread breaks down at constriction and mercury fills the bulb and a part of capillary. An ordinary mercury thermometer, called auxiliary thermometer, is attached on one side of reversing thermometer.

Method

Lower the reversing thermometer units to the desired depth waits for a few minutes and reverse the thermometer with the help of a messenger. Take it out and note the readings in two thermometers.

Calculation:

$$T = T_1 + T'$$

$$T = \frac{(T_1 - t)(T_1 - V_0)}{K - 100} + I$$

Where, T = corrected temperature of reversing thermometer (°C); T_1= uncorrected reading of reversing thermometer; t= temperature recorded from auxiliary thermometer; V_0= volume of the small bulb and of capillary up to 0° C graduation; K= a constant (= 6100) depending upon the relative expansion of mercury and the type of glass used in thermometer; I= calibration correction for reversing thermometer.

C- Other methods

Other methods used for subsurface temperature recording are the thermistor, thermo phone, and bathythermograph etc.

Annual heat budget of a lake is the amount of heat required to raise its water from minimum winter temperature to maximum summer temperature.

Requirements:

(i)   Thermometer

(ii)   Depth gauge- A weight tide to a graduated rope.

Method

Record, round the year, the temperature at various levels of body of water, and its depth. From these data calculate the mean summer temperature, mean winter temperature, and mean depth. Calculate annual heat budget of tropical and subtropical waters as follows:

Calculation

$$\text{AHB} = Z^S \left(T_m^s - T_m^w \right) + \left(Z^w - Z^s \right) \times 470$$

Where, AHB = Annual heat budget (cal/cm^2/year); Z^s = mean depth during summer (cm); Z^w = mean depth during winter (cm); T_m^s = mean temperature during summer (^{o}C); T_m^w = mean temperature during winter (^{o}C).

470 is the factor computed from latent heat of vaporization (i.e., 540 cal) minus work done against external pressure (i.e., approx. 70 cal.).

Mean temperature (during summer or winter) is calculated as follows:

$$T_m = \frac{T_1 + T_2}{2}(P_1) + \frac{T_2 + T_3}{2}(P_2) + ----- \frac{T_n-1 + T_n}{2}(P_n)$$

Where, T_m = mean temperature; T_1, T_2, T_3 = Temperatures at respective levels bounding the various strata of water from surface to downward; T_n = Temperature at lower most level of last stratum; P_1, P_2, P_n = percentage of entire lake volume which the first, second, and the last strata respectively constitute (e.g., if one stratum constitutes 10% of the entire lake volume, then the P = 10/100 = 0.1).

Introduction

The hydrogen ion concentration or pH is defined as the negative base 10 logarithm of the hydrogen ion concentration in moles per liter. It is the intensity of acidity or alkalinity. The pH scale ranges from 0- 14 with 7 being a neutral value. Pure water has a pH of 7 which means it 1×10^{-7} moles per liter of hydrogen ions. pH is a significant factor which determines the strength of water for various purposes. pH of natural waters varies around 7, generally over 7(i.e., alkaline) due to presence of sufficient quantities of carbonates. It increases during day largely due to photosynthetic activity (consumption of carbon di oxide), whereas decreases at night due to respiration.

The variations in pH are linked with the life processes of inhabitant organisms. It is also contributing to the productivity of ecosystems by influencing the species composition and affects the availability of nutrients and the relative toxicity of many trace elements. The pH in rivers and lakes is affected by the geology of the water source, atmospheric inputs and a range of other chemical components.

The pH is the negative logarithm of hydrogen ion concentration. It is expressed as pH = -log (H) $^+$.

Water does not show any reaction with acid, when the dissociation product of it are in equilibrium ($HOH \rightarrow H^+ + OH^-$, $H^+ = OH^- = 10^{-7}$). pH at this point is 7. When H^+ concentration increases, water becomes acidic and when decreases it becomes alkaline.

pH scale :   0 _____ 7 _____ 14.

                        Acidic   Neutral   Alkaline.

The pH meter is the most widely used electrical method for finding out the hydrogen ion concentration of a sample.

Electrometric measurement of pH

In this method a Glass Electrode and a Reference Electrode are inserted in a solution and the electrical potential or voltage across these electrodes is taken as a measure of the hydrogen ion in the solution. Reference Electrode is generally the Calomel Electrode.

With ordinary Glass Electrode pH can be measured in a range of 2 to 10 pH only as the Glass is made up of sodium silicate which gets affected by pH beyond this range. Special Glass Electrode called the universal Glass Electrode is used to detect pH in full range of 0 to14.

Measurement of pH using pH meter:

Requirement:

(i)   Combined electrode. Portable field pH meter which has a single electrode and is convenient to use.

(ii)   Thermometer

(iii)   Buffer solution, of pH 4.0 and 9.2. Buffers of different pH can be prepared by dissolving standard pH tablets in 100 ml water, or as follows:

(a) Phthalate buffer: Dissolve 10.2 g of potassium hydrogen phthalate in distilled water to make 1 liter of buffer.

(b) Phosphate buffer: Dissolve 3.40 g of KH_2PO_4 and 4.45 g of Na_2HPO_4 $2H_2O$ in distilled water to make 1 liter of buffer.

(c) Borax buffer: Dissolve 3.81 g of $Na_2B_4O_7$. 10 H_2O in distilled water to make 1 liter of buffer.

Method:

It's better to read operation manual of pH meter before using it. Put the switch on and see that the meter reads pH 7. Wash the electrodes with distilled water and connect it with the pH meter. Dip the electrode in buffer of pH 4.0 and move the temperature compensation knob to the temperature of buffer. Turn selector switch to pH range of 0-7 and adjust the set buffer knob until the meter reads pH 4.0. Move selector switch to zero, remove the electrode from buffer, wash it with distilled water, and dip in a buffer of pH 9.2. Turn the selector switch the pH range of 7 to 14 and adjust the set buffer knob until the meter reads the pH 9.2. In doing so, the pH meter is calibrated both for pH ranges from 0 to 7 and from 7 to 14. Put selector switch to zero. Wash the electrode with distilled water and dip in the sample. Adjust temperature compensation knob to the temperature of the sample. Put the selector switch to pH range of 0 to 7 and read the meter for the pH of the sample. If pH exceeds 7, move selector switch to pH range of 7 to 14 and read the meter. Turn the selector switch to zero, switch off the instrument, and remove the electrode. Keep the electrode dipped in distilled water when not in use.

Indicator Method for the measurement of pH:

Series of indicator and buffer solutions is required for indicator method. A universal indicator is prepared by adding 0.05 g methyl orange, 0.15 g methyl red, 0.3 g bromothymol blue, and 0.35 g of phenolphthalein in 1 liter of 66% alcohol.

Color changes shown by universal indicator are as follows:

pH		Color
Up to 3	:	Red
4	:	Orange red
5	:	Orange
6	:	Yellow
7	:	Yellowish green
8	:	Greenish blue
9	:	Blue
10	:	Violet
11	:	Reddish violet

Table-4: Individual Indicators for Various pH Ranges:

S. No.	Name of Indicator	pH Range	Color Change
1	Thymol blue (acid range)	1.2 to 2.8	Red to yellow
2	Bromophenol blue	3.0 to 4.6	Yellow to blue violet
3	Bromocresol Green	3.8 to 5.4	Yellow to blue
4	Methyl red	4.2 to 6.3	Red to yellow
5	Bromocresol Purple	5.2 to 6.8	Yellow to blue violet
6	Bromothymol blue	6.0 to 7.6	Yellow to blue
7	Phenol red	6.8 to 8.4	Yellow to red
8	Cresol red	7.2 to 8.8	Yellow to red
9	Thymol blue (alkali orange)	8.0 to 9.6	Yellow to blue
10	Thymolphthalein	9.3 to 10.5	Colorless to blue
11	Thymol violet	9.0 to 13.0	Yellow to green to violet

Total Dissolved Solids:

A large number of salts are found dissolved in natural waters, the common of them are carbonates, bicarbonates, chlorides, sulfates, phosphates, and nitrates of potassium, calcium, magnesium, iron, potassium, sodium, and manganese etc. High content of dissolved solids elevates the density of water, influences osmoregulation, of fresh water organisms, reduces solubility of oxygen, and utility of water for drinking and cultures, irrigation and industrial purposes. It is an important parameter in the analysis of saline lake, estuaries, coastal, marine, and fresh water as well. This expressed in terms of g/l or ppt (parts per thousands).

Requirements

Evaporating dish, chemical valence, desiccator, hot water bath, and Whatmann filter paper No 4.

Method

Take an evaporating dish of suitable size, make t clean and dry, and weigh it. Filter 250 to 500 ml of sample through a whatmann filter paper No. 4. Take the filtrate in evaporating dish. Evaporate the sample on hot water bath. When whole water is evaporated note the weight of evaporating dish after cooling it in a desiccator.

Calculation: $\text{TDS (g/l)} = \left[\dfrac{A-B}{V}\right] \times 1000$

Where, TDS= Total dissolved solids; A= Final weight of evaporating dish (g); B = Initial weight of evaporating dish (g); and V = Volume of sample taken (ml).

Dissolved solids:

Pipette out 100 ml of the well mixed sample and filter through a filter paper. Collect the filtrate in a clear and previously weighed evaporating dish at room temperature and weigh.

$$\text{Total dissolved solids mg/l} = \frac{\text{mg of residue} \times 1000}{\text{ml sample}}$$

Total solids:

To determine total solids no need to filter the sample. Pipette out 100 ml of well mixed sample in to the clear dry weighing dish and evaporate to dryness at 105°C. Cool the weighing dish at R.T. and weigh.

$$\text{Total Solids mg/l} = \frac{\text{mg of residue} \times 1000}{\text{ml sample}}$$

Fixed Residue: Ignite the residue obtained in total solids determination for volatile solids determination at 550° C to 600° C for 1 hour in the platinum crucible in the muffle furnace. Cool the crucible and weigh. Find out the amount of fixed residue in the crucible. This residue is known as fixed residue and contains dissolved non volatiles as well as suspended nonvolatile residue. This is also known as ignited residue also.

$$\text{Ignited residue or fixed residue mg/l} = \frac{\text{mg or residue} \times 1000}{\text{ml sample}}$$

$$\text{Volatile residue} = \text{Total solids} - \text{Ignited solids}$$

Suspended matter: It is determined as follows:

$$\text{Suspended matter mg/l} = \frac{\text{Total solids mg/l}}{\text{Dissolved solids mg/l}}$$

Or, if Glass Fiber Filter paper GFC grade (Whatmann make) is available this determination can be done directly. This paper keeps the consistency not only at 105° C but at 600° C also. While using this filter paper a separate assembly known as Hartley's Filter Assembly is used.

Introduction

Alkalinity can be defined as the "capacity of the sample to neutralize an acid of known strength". Alkalinity is caused by hydroxides, carbonates, bicarbonates, phosphates, silicates, salts of weak acids, weak or strong factors base, etc. It has buffering capacity in accommodating H^+ and OH^- ions in order to maintain the neutrality. Alkalinity, hardness and pH affect the toxicity of many substances.   It has buffering capacity in accommodating H+ and OH- ions in order to maintain the neutrality. It is expressed as mg/l $CaCO_3$.

Requirements:

Glassware's- Burette, pipette, beaker

Indicators- Phenolphthalein indicator: End point –Red to colorless.

Methyl orange indicator: End point – Yellow to orange.

Titrant: N/50 of 0.02 N H_2SO_4.

Procedure:

Add phenolphthalein indicator to 50 ml Aliquot and titrate with 0.02 N H2SO4 so as to get color less solutions. Note the reading. Add methyl orange to the same sample and titrate till orange color appears. Note the readings.

1 ml of 1 N. H_2SO_4 = 50mg $CaCO_3$.

1 ml of 0.02 N. H_2SO_4 = 1 mg $CaCO_3$.

Calculations:

Phenolphthalein alkalinity ('P' alkalinity)

$$\text{mg/l } CaCO_3 = \frac{\text{Volume of } H_2SO_4 \text{ required to get end point} \times 1000}{\text{Volume of sample}}$$

Methyl orange Alkalinity 'M' or 'T' Alkalinity

$$\text{mg/l } CaCO_3 = \frac{\text{Volume of 0.02 N. } H_2SO_4 \text{ to get end point} \times 1000}{\text{Volume of sample}}$$

Introduction

Pure water is poor conductor of electricity. Acids, bases, and salts in water make it relatively good conductor of electricity, such substances are called electrolytes. On the bases of solubility of electrolytes in water, they are differentiated into weak and strong electrolytes. The electrolytes highly soluble in water are known to be as strong electrolytes, and those which are less soluble in water are called as weak electrolytes. In other words, electrolytes in a solution dissociate into positive and negative ions, which are called cation's and anion's respectively, and impart conductivity. Thus, higher the concentration of electrolytes in water the more is its electrical conductivity (i.e., lesser the resistance). Conductance is the reciprocal of resistance involved and the unit of measure of conductance is reciprocal ohm designated as mho or Siemens (S).

Determination of electrical conductivity provides a rapid and convenient means of estimating the concentration of total dissolved solids in water. If conductivity and dissolved solids correlation factor is found out previously ; dissolved solids need not be determined in future as it can be computed by multiplying conductivity with the correlation factor. With water samples of variable proportions of neutral salts the factor also varies. Conductivity measurements are made after proper temperature compensation. It is claimed that, other things being equal, the richer a body of water in electrolytes the greater is its biological productivity.

Conductivity = Dial reading ×Multiplier reading micro mhos

Specific Conductivity =

Dial reading ×multiplier reading ×cell constant micro mhos

Cell Constant Check = Take a solution of known conductivity and 0.01 N KCl of reagent grade to compensate for temperature and measure the conductivity.

$$\text{Cell Constant} = \frac{\text{Specific Conductivity of the solution}}{\text{Measured conductivity}}$$

At 25° C, the specific conductivity of N/100 KCl solution is 140 micro mhos. This solution is suitable for cell constant between 0.1 to 10.

Introduction

This is due to the presence of divalent metallic cations, e.g., Ca^{2+}, Mg^{2+}, Fe^{2+}, and $Sr2^{+}$ etc. Water hardness is important to fish culture and is a commonly reported aspect of water quality. It is a measure of the quantity of divalent ions such as calcium and magnesium in water. These divalent ions are the most common source of water hardness. Calcium has an important role in the biological processes like bone formation, blood clotting, and other metabolic reactions. Fish can absorb calcium directly from water and food.

Requirements

Glass wares- Burette, Pipette, beaker, conical flask, etc. Chemicals/ reagents used:

a- Buffer solution: It is prepared by dissolving 16.90 gm. NH_4Cl in 143 ml. concentrated NH_4OH and then adding 1.25 gm. of magnesium salt of EDTA and dilute to 250 ml with H_2O.

b- Eriochrome Black T indicator: Mix together 0.5 g. Eriochrome Black T and 100 g NaCl.

c- Standard EDTA titrant, 0.01 M.: Dissolve 3.723 g disodium salt of EDTA in 1 ml H_2O. Check the titer by standardizing against standard $CaCO_3$ solution

1ml. = 1 mg. as $CaCO_3$.

d- Standard $CaCO_3$ solution: 1 gm. anhydrous $CaCO_3$ dissolve in 1 ml HCl and add 200 ml H_2O. Boil to expel CO_2. Cool and add Methyl Red indicator and adjust to the intermediate orange color by adding 3 N. NH_4OH or HCl. Dilute to 1 liter.

Procedure

Dilute 25 ml of sample to about 50 ml with H_2O. Add 1-2 ml of buffer solution. Add about 0.2 gm. dry powder indicator formulation. Titrate with standard EDTA solution. At the end point color changes from wine red to blue.

Calculation

Hardness (EDTA) as mg/l $CaCO_3$ = $\dfrac{A \times B \times 1000}{\text{ml. Sample.}}$

Where, A = ml titration for sample and, B = mg $CaCO_3$ equivalent to 1.0 ml titrant.

(i) Calcium Hardness:

Principle:

When EDTA is added to water containing both calcium and magnesium, it combines first with the calcium. Calcium can be determined directly using EDTA when the pH is made sufficiently high so that the magnesium is largely precipitated as the hydroxide and an indicator is used which combines with calcium only. NaOH indicator Ca chelate Murexide indicator (purple) + Ca^{2+} (Part) pH >10

Process: Ca^{2+} + EDTA − EDTA - Ca chelate

EDTA + Indicator – Ca chelate = EDTA-Ca chelate + Indicator

Reagents:

(a) NaOH 1 N: Dissolve 40 g NaOH in 1 liter H_2O.

(b) Murexide (amm. purpurate) indicator: Mix 0.2 g of murexide with 100 g NaCl in a grinder.

(c) Standard EDTA

Procedure

Add 2.0 ml NaOH and 2.0 g of indicator to the 50 ml of water sample. Titrate this solution with standard EDTA. At the end point color changes from pink to purple or violet.

Calculation:

$$\text{mg/l Ca} = \frac{A \times B \times 400}{\text{ml sample}}$$

$$\text{Ca Hardness as mg/l CaCO}_3 = \frac{A \times B \times 1000}{\text{ml sample}}$$

Where, A= ml titration for sample and B= mg $CaCO_3$ equivalent to 1.00 ml EDTA titrant at the calcium indicator end point.

Introduction

Sulfates occurs widely in everyday life, they are found as microscopic particles (aerosols) resulting from fossil fuels and biomass combustions. Sulfates are used as gypsum and copper sulfate in forms and fields as fertilizer and algaecide. It is also used in soaps, shampoo, medicines and electrical. In excess amounts sulfate increases acidic condition in environment and form acid rains

(i)   Titrimetric analysis

Reagents used:

    a- EDTA Solution: 4 gm. disodium di-hydrogen ethylene di-amine tetra acetic acid dehydrate are dissolved in one liter of distilled water. Standardized against standard $BaCl_2$ solution.

    b- Buffer: Mix 67.5 gm. pure NH_4Cl with 750 ml NH_4OH (special grade 0.92) and dilute to 1 liter.

    c-   Indicator: Erichrome Black T.

    d- $BaCl_2$ standard: Dissolve 2.443 gm. $BaCl_2$ $2H_2O$ in 1 liter. 1 ml solution = 1 mg $CaCO_3$ or 0.96 mg $-2SO_4$.

Procedure

Neutralize 100 ml with 1 $NHNO_3$. Boil to expel CO_2. Add 10 ml or more if required, of $BaCl_2$ standard solution to the boiling sample and allow cooling. Dilute to 200 ml, mix and allow the precipitate to settle. Withdraw 50 ml supernatant; add 0.5 to 1 ml buffer and indicator. Titrate with standard EDTA till blue color appears.

$$-2SO_4 \text{ mg/l} = 9.6 [(0.1A+B)-4C]$$

Where,

A= Total hardness of the sample as $CaCO_3$ mg/l.

B= Volume of $BaCl_2$ in ml added.

C= Volume of EDTA required for titration.

(ii) Turbidimetric Analysis:

Principle: Sulfate is precipitated as barium sulfate in hydrochloric acid medium with barium chloride, in a near colloidal form. The absorbance of light caused by the barium sulfate suspension is measured by a nephelometer or colorimeter and the sulphate ion concentration is determined by using a calibration curve prepared with standard solution.

Requirements:

Magnetic stirrer, Klett Summerson colorimeter or spectrophotometer, Measuring Spoon

Reagents;

(a) Conditioning reagent- Mix 50 ml solution with a solution containing 10 ml concentrated HCl, 300 ml distilled water , 100 ml 95% ethyl alcohol and 75 g sodium chloride.

(b) Barium chloride crystals 'AR grade'

(c) Standard sulfate solution- Dilute 10.41 ml of standard 0.02 N H_2SO_4 to 100 ml with distilled water.

Procedure:

1- Formation of barium sulfate turbidity:

Measure 100 ml sample or a suitable aliquot made up to 100 ml into a 250 ml Erlenmeyer flask. Add exactly 5 ml conditioning reagent and mix in the stirring apparatus. While the solution is being stirred add a spoon full of barium chloride crystals. Stir for exactly 1 minute at a constant speed.

2- Measurement of barium sulfate turbidity;

Immediately after stirring, pour some of the solution into the absorption cell of the photometer and measure the absorption at fifth minute. Maximum turbidity is usually achieved within 2 minutes, and the reading remains constant thereafter for 10 minutes.

Calculation: $\text{mg } SO_4^{2-} = \dfrac{\text{mg } SO_4^{2} \times 1000}{\text{ml sample}}$

Conditioning solution

Reaction: $SO_4^{2-} + Ba\,Cl_2 \xrightarrow{\hspace{3cm}} BaSO_4$

Introduction

Chloride ions are important compounds of all living system contributing to the osmotic, ionic as well as water regulation functions within organisms. Chloride occurs naturally in all type of water in variable concentrations. The concentration values of chloride considered safe by WHO is 200 mg/l. above this concentration it produces salty taste. Chloride concentration serves as an indicator of pollution by sewage. The ecological significance of chloride lies in its potential to regulate salinity of water and exert consequent osmotic stress on biotic communities.

Titrimetric Analysis

In neutral or slightly alkaline solutions K_2CrO_4 indicates the end point of the $AgNO_3$ titration with chlorides. Silver chloride gets quantitatively precipitated before red silver chromate is formed.

Reagents used:

a- Chloride free water: Deionized water.

b- K_2CrO_4 indicator: Dissolve 50 gm. in distilled water. Add $AgNO_3$ solution till a definite red color precipitate is formed. Allow it to stand for 12 hours and filter. Dilute the filtrate to1 liter.

c- Standard $AgNO_3$(0.0141 N.): Dissolve 2.395 gm. $AgNO_3$ in distilled water and dilute to 1 liter 1 ml of standard silver nitrate (0.0141 N) is equivalent to 500 micro gram Cl.

d- Standard NaCl (0.0141 N.): Dissolve 824.1 mg NaCl in chloride free water and dilute to 1 liter. 1 ml of this solution is equivalent to 500 micro liters Cl.

Procedure:

1- Use 100 ml of suitable aliquot diluted to 100 ml with distilled water.

2- If the sample is colored, add 3 ml Al (OH)$_3$ suspension. Mix, allow settling, filtering wash and combining the filtrate and washings.

3- Check the pH and adjust it to near neutrality.

4- If sulphide, sulphate, or thio-sulphate is present, make the water alkaline to phenolphthalein with NaOH. Add 1 ml H$_2$O$_2$ stir, neutralize with H$_2$SO$_4$.

5- Titrate the sample with AgNO$_3$ after adding 1 ml. K$_2$CrO$_4$ to the sample till orange red color appears.

6- Run the blank taking distilled water as the sample.

Calculations:

1000 ml of 1 N. AgNO$_3$ = 35.45 g Cl$^-$.

Therefore,

1 ml of 0.0141 N. AgNO$_3$ = 35.45 * 0.0141 mg Cl-.

= 0.499 mg Cl- or say 500 micro grams Cl- or 0.5 mili grams Cl-.

Therefore,

$$\text{mg/l Cl-} = \frac{(A-B) \times 0.5 \times 1000}{\text{ml. of sample.}}$$

Where;

A = Volume of titrant for sample, B = Volume of titrant for blank.

45

Introduction

Nitrogen estimation is of significance to sanitary engineering practices, prior to the knowledge of bacteriological procedures. Determination of various forms of nitrogen on water was done to assess its quality. Presence of organic and ammonia nitrogen or total nitrogen is taken as a chemical evidence of the oxidence of recent organic pollution, particularly of animal origin. On the other hand, presence of the oxidized forms, viz. NO2, NO3 used to indicate remote pollution. Ammonia above a certain level is toxic to fish life in surface waters.

Nitrates in drinking water have been reported to be a cause of disease Methaemoglobinemia. However, presence of nitrogen in appropriate amount in waste water is necessary for its treatment through biological processes. Acknowledge of nitrogen in waste waters is important when it is used as an irritant.

A- Ammonia Nitrogen

Principle:

Ammonia produces a yellow colored compound with alkaline Nessler reagent, provided sample is clear. Pretreatment with $ZnSO_4$ and $NaOH$ precipitates Ca, Fe, Mg sulfide removes Rochelle salt solution prevents precipitation or residual Ca and Mg in the presence of alkaline Nesslers reagent.

Interference;

Color, turbidity, Ca, Mg, salts and Fe in the sample constitute the price sources of interference.

Requirements:

Spectrophotometer having range of 300-700 nm and Nessler tubes or volumetric flasks of 100 ml capacity.

Reagents:

    (a) Zinc sulfate: Dissolve 10 g of $ZnSO_4H_2O$ in distilled water and dilute to 100 ml.

    (b) Sodium hydroxide 6 N: Dissolve 24 g NaOH and dilute to 100 ml.

    (c) EDTA: Dissolve 50 g EDTA in 60 ml water containing 10 g NaOH. Cool and dilute to 100 ml.

    (d) Rochelle salt solution: Dissolve 50 g potassium sodium tartrate tetra-hydrate $KNaC_4^-H_4^-O_6 \cdot 4H_2O$ in 100 ml water. Remove ammonia by boiling off 30 ml solution, cool and dilute to 100 ml.

    (e) Nesslers reagent: Mix well 100 g $HgCl_2$ and 70 g KI and dissolve small quantity of water. Add this mixture to a cooled solution of 160 g NaOH in 500 ml, and dilute to 1000 ml. Keep overnight. Store supernatant in colored bottle and away from sunlight (Take care to avoid ingestion since it is toxic).

    (f) Standard ammonia solution: Dissolve 3.819 g anhydrous NH_4Cl dried at 100° C in distilled water and dilute to 100 ml. Dilute 10 ml of the solution to 1000 ml. 1 ml = 10µg N or 12.2 µg NH_3.

Procedure:

1- Take 100 ml of sample and add 1 ml $ZnSO_4$ solution, and 0.5 ml NaOH to it so as to obtain a pH of 10.5. Allow to settle and filter the supernatant through 42 no. Whatmann filter paper.

2- Take a suitable aliquot of sample and dilute to 50 ml.

3- Add 3 drops of Rochelle salt solution or 1 drop of EDTA, and mix well.

4- Add 2 ml. Nesslers reagent if EDTA is used or 1 ml if Rochelle is used. Make up to 100 ml.

5- Mix well, and after 10 minutes, read percent transmission at 410 μμ or blue filter (42), using blank prepared in same way using distilled water instead of sample. Prepare a calibration curve suitable aliquot of standard solution in the range of 5 to 120 μg/ 100 ml for reference following the same procedure as to 1 to 5 but using the standard solution in place of sample.

Reaction:

a- Ammonia- N

Nitrogen that exists as NH_4^+ or in the equilibrium:

$$NH_4^+ \xrightarrow[\text{Acidic pH}]{\text{Alkaline pH}} NH_3 + H^+$$

is considered to be as ammonia nitrogen.

(i) Nesslerization: $NH_3 + 2\ K_2HgI_4 + 3KOH^-$

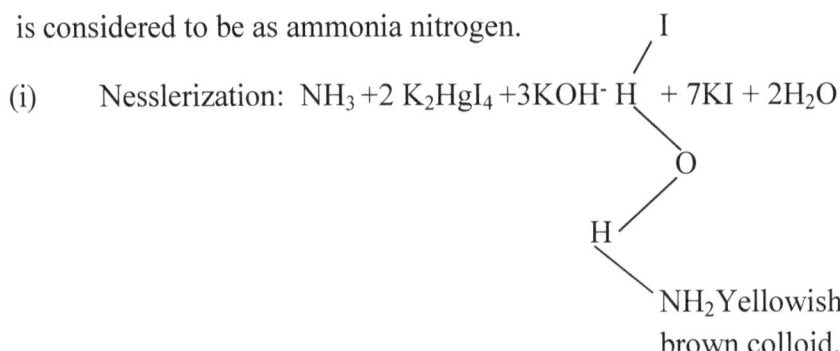

$+ 7KI + 2H_2O$

NH_2 Yellowish brown colloid.

Color is matched with the standard colors produced by concentrations of NH_4^+ visually or on photometer using 420 mμ filters.

(ii) Distillation: $NH_4^+ \xrightarrow[\text{Heat and alkali}]{} NH_3 + H^+$

$NH_3 + H_3BO_3 \xrightarrow{\text{Mixed indicator}} NH_4^+ + H_2BO_3^-$

pH = 9.2 (blue color)

$H^+ + H_2BO_3 \longrightarrow H_3BO_3$ (Red color pH – 4.8)

b- Kjeldahl Nitrogen

Principle:

In the presence of sulfuric acid, potassium sulfate, mercuric sulfate, catalyst organic nitrogen gets converted to ammonium sulfate. Potassium sulfate is added to raise the boiling point of sulfuric acid from $345°$ to $375°$ C. The digest is diluted and NaOH is added to make it alkaline, then it is distilled. The liberated ammonia is absorbed in boric acid. The absorbed ammonia is determined either by titration or direct nesslerization. Temperature should be kept below $382°$ C during digestion to avoid loss of nitrogen.

Interference:

Sludge solids, if present in large quantity, bump and result in loss of nitrogen. To prevent bumping a little amount of paraffin wax is added during digestion. It should be kept in mind that in such cases digestion acid should be increased to keep a ratio of salt to acid as 0:8.

Reagent:

(i)   Concentrated Sulfuric acid AR Grade

(ii)   Potassium per sulfate crystals

(iii)   Boric acid AR : 2 percent solution

(iv)   Mixed indicator: 100 mg Bromo-cresol-green in 100 ml. Ethyl alcohol and 20 ml methyl red in 20 ml ethyl alcohol mix well.

(v)   12 N NaOH

(vi)   0.02 N H_2SO_4

Digestion and Distillation in Kjeldahl Apparatus:

Take a suitable aliquot of sample in a digestion flask and add to it 20 ml or more concentrated sulfuric acid and potassium pre-sulfate crystals as catalyst. Boil the mixture till it becomes clear or pale straw colored. Dilute it to a definite volume in volumetric flask. Take a suitable aliquot and add it to a kjeldahl distillation flask. Add 12 N NaOH to make the mixture alkaline. Distill it with steam so as to liberate ammonia. Collect and trap the ammonia vapors in boric acid with mixed indicator (red colored) till the volume in flask becomes doubled and the color is turned to blue. Remove the flask and titrate the contents with 0.02 N sulfuric acid till the original red color is obtained.

Calculation:

mg nitrogen present in the sample taken for the digestion =

$$\frac{ml\ 0.02\ N\ H_2SO_4 \times 0.28}{ml\ sample\ taken\ for\ distillation} \times Volume\ of\ digest\ after\ dilution$$

B- Nitrite-

Occurrence: Trace amount of nitrite may indicate organic pollution. Water containing nitrate acquire nitrite by reduction when standing in contact with iron or other metals.

Principle: Sulphanilic acid is diazotized with nitrite in acid solution and the resulting diazo compound is coupled with 1-naphthylamine hydrochloride to form a purplish-pink azo dye.

Sulphonilic acid   **Diazonium salt**

α- Naphthylamine   Red colored azo dye

Interference: Amines, strong reducing agents and strong oxidizing agents are the only substances known to interfere.

Minimum detectable concentration: 1 μg/l.

Storage of sample: Determination should be made promptly.

Reagents:

(1) Sulphanilic acid reagent: Dissolve 0.6 g sulphanilic acid in 70 ml hot distilled water, cool, add 20 ml concentrated HCl dilute to 100 ml with H_2O and mix thoroughly.

(2) 1-Naphthalamine hydrochloride reagent: Dissolve 0.6 gm 1-naphthalamine hydrochloride in H_2O, to which 1 ml concentrated HCL has already been added. Dilute to 100 ml with H_2O and mix thoroughly.

(3) Sodium acetate buffer solution, 2 M: Dissolve 16.4 g $NaC_2H_3O_2$ or 27.2 g of $NaC_2H_3O_2$. $3H_2O$ in H_2O and make up to 100 ml.

(4) Nitrite Stock Solution; Dissolve 150 mg $NaNO_3$ in water and dilute to 1 liter.

$$1.0 \text{ ml} = 100 \text{ μg nitrite as } NO_2.$$

(5) Nitrite working solution: Dilute 10 ml stock solution to 10. The solution must be freshly prepared.

1 ml = 1.0 μg nitrite as NO_2.

(6) EDTA Solution: Dissolve 0.5 g sodium salt of EDTA in 100 ml H_2O.

Procedure: Add 1 ml EDTA solution and 1 ml sulphanilic acid reagent to 50 ml clear water sample. After 3-10 minutes add 1ml naphthylamine hydrochloride reagent and 1 ml sodium acetate buffer solution and mix well. At this point the pH of the solution should be 2.0 to 2.5. After 10 to 30 minutes, measure the absorbance at 540 μu against a reagent blank.

Prepare standards in nessler tubes by adding the following volumes of standard $NaNO_2$ solution and diluting to 50 ml with H_2O:

0, 0.1, 0.2, 0.4, 0.7, 1.0, 1.2, and 1.5 ml.

Calculation: $$\text{mg/l nitrite N} = \frac{\mu g \text{ nitrite N}}{\text{ml sample}}$$

mg/l NO_2 = mg/l nitrite N × 3.29

C- Nitrate

Introduction

Nitrate is the highest oxidized form of nitrogen and in water its most important source is biological oxidation of nitrogenous organic matter of both autochthonous and allochthonous origin. The high concentration of nitrates in water is indicative of pollution. This is an important plant nutrient, when present in excess; it causes ubiquitous growth of algae, often present in blooms. High nitrate content (>40 mg NO_3-N/l) may cause blue-baby disease.

Nitrate (NO_3^-) is a major nutrient essential for plant growth. Nitrogen may be present in water as dissolved nitrogen gas, organic nitrogen compounds and inorganic nitrogen compounds including ammonia, nitrite and nitrate.

Sources of nitrate nitrogen include precipitation, nitrogen fixation in water and sediments by bacteria and certain cyanobacteria, and inputs from surface and ground water drainage. Agriculture is the primary source for excess nitrate. Excess nitrate fertilizers from agricultural area gets into the water sheds through surface run off. On the other hand nitrate is also a by-product of septic systems. As such, water quality may also be affected through ground water flow paths in water sheds that have a high number of septic systems.

Spectrophotometric Analysis

Requirements: 1- Spectrophotometer; 2- laboratory glassware; 3- hot water bath; and 4-chemical reagents as described below:

a- Phenol di-sulfonic acid: Dissolve 25 grams of white phenol in 150 ml of concentrated sulfuric acid and further add 85 ml of concentrated sulfuric acid. Heat for about 2 hours on a water bath, cool, and keep the solution in a dark bottle.

b- Potassium hydroxide solution (12 N): Dissolve 336.5 grams of potassium hydroxide in distilled water to make the volume 500 ml.

c- Standard nitrate solutions: Dissolve 0.722 gram of anhydrous potassium nitrate in distilled water to prepare 1 liter of stock solution. This stock solution contains 100 mg NO_3-N/l(or 443 mg NO_3 ions/ l).

d- Prepare standard nitrate solutions of various strengths (preferably in the range from 0.0 to1.0 mg NO_3-N/l at the intervals of 0.1 mg NO-N/l) by diluting stock solution with distilled water.

Method:

Take 25 ml of sample in a porcelain basin and evaporate it to dryness on a hot water bath. Add 0.5 ml of phenol disulphonic acid (reagent A) to the residue and dissolve the latter with the help of a glass spatula. Add 5 ml of distilled water and 1.5 ml of potassium hydroxide solution (reagent B). Stir for thorough mixing. Take the supernatant of yellow color read its absorbance(S) on spectrophotometer at 410 nm. Distilled water is used as blank.

Process the standard nitrate solutions (reagent C) in similar manner and note the absorbance for each. Plot a standard curve between absorbance and concentrations of various standard solutions.

Deduce the value of nitrate and nitrogen in the sample by comparing the absorbance of sample (S) with the standard curve and express the result in mg NO_3-N/l.

Phenol-di-sulfonic acid Method:

Principle: The yellow color produced by the reaction nitrated and phenol-di-sulfonic acid obeys Beers law up to at last 55 mg/l NO_3 at a wavelength of 410 nm when a light path of 1 cm is used.

Interference: Chloride concentration above 10 mg/l.

Apparatus:

   (a) Spectrophotometer: For use at 410 m, providing a light path of 1 cm or longer.

   (b) Nessler tubes: Matched, 50 100 ml.

Reagent:

(a) Standard Ag_2SO_4 solution: Dissolve 4.40 g Ag_2SO_4 in distilled water.

   1 ml= 1.00 mg Cl.

(b) Phenol-di-sulfonic acid reagent: Dissolve 25 g phenol in 150 ml concentrated H_2SO_4, stir and heat for 2 hours on a hot water bath.

(c) Ammonium hydroxide concentrated.

(d) Potassium hydroxide solution (12 N KOH): Dissolve 673g KOH in distilled water and dilute to 1 liter.

(e) Stock nitrate solution: Dissolve 1.6308 g KNO_3 in 1 liter H_2O.

   1 ml of this solution = 1 mg NO_3.

(f) Standard nitrate solution: Evaporate 50 ml of KNO_3 stock solution to dryness on a water bath, dissolve the residue by rubbing with 2 ml of phenol-di-sulfonic acid reagent, and dilute to 500 ml with H_2O.

   1 ml of this solution = 0.1 mg NO_3

   1 ml/ 50 ml = 2.0 mg/l NO_3.

(g) Aluminium hydroxide suspension: Dissolve 125 g of potassium or ammonium alum in 1 liter H_2O. Warm to 60°C and add 55 ml of concentrated NH_4OH slowly and continuous stirring. After 1 hour wash the precipitate by distilled water until free from ammonia, nitrate, nitrite, and chloride.

Reaction:

$$C_6H_5OH + 2H_2SO_4 = C_6H_3 (OH) (SO_3H)_2 = 2H_2O$$

Phenol di sulfonic acid

$$C_6H_3 (OH)(SO_3H)_2 + HNO_3 = C_6H_3(OH)(SO_3H)_2NO_2 + H_2O$$

Colorless compound

$\downarrow$ KOH

$$C_6H_3O (SO_3K)_2 NO_2 K$$

A yellow colored salt.

Procedure:

(a) Preparation of standards:

Introducing into a series of 50 ml Nessler tubes the following volumes of standard KNO_3 solution:

0, 1.0, 2.5, 5.0, 7.5, 10.0, 12.5, 15, 17.5, 20, 22.5, 25, 27.5 (i.e. from 0-55 mg/l) and add 2 ml of phenol-di-sulfonic acid and a volume of the alkali (concentrated NH_4OH or KOH) equal to that used for the samples. Make up to 50 ml, shake. These standard solutions may be kept for several weeks without deterioration.

(b) Color removal: Add 3 ml aluminium hydroxide suspension to 150 ml colored water sample, then filter it.

(c) Nitrite conversion

(d) Chloride removal: Treat 100 ml of sample with an equivalent amount of standard silver sulfate solution for chloride removal. Remove the precipitate of silver chlorides by filtration.

(e) Neutralize the clarified sample to approx. pH 7. Transfer to a casserole, and evaporate to dryness over a hot water bath. Rub residue with 2.0 ml phenol-di-sulfonic acid using a glass rod. Dilute with 20 ml H_2O and add with stirring about 6 to 7 ml NH_4OH or with 5 to 6 ml KOH until maximum color is developed. Transfer to 50 ml Nessler tube.

Make photometric reading of the standards and the sample with a 1 cm or longer light path at a wavelength of 410 nm.

Chapter–16   **Phosphates**

Introduction

Phosphorus is the first limiting nutrient for plant growth in fresh water, which regulates the production of phytoplankton's in presence of nitrogen. It is available in the form of phosphate (PO_4) in natural waters and generally occurs in low or moderate concentrations. Agricultural runoff containing phosphorylated compounds like insecticides, and fertilizers as well as waste waters containing detergents etc. tend to increase phosphate pollution in water. Inorganic phosphorus plays a dynamic role in ecosystems; when present in low concentration is one of the most important nutrients, but in excess along with nitrate and potassium, causes eutrophication leading to algal growth called algal blooms.

Spectrophotometric Analysis

Requirement:

1- Spectrophotometer; 2-Laboratory glassware; 3- hot plate; and 4- chemical reagents, as given below:

a- Per chloric acid (70%),

b- Phenolphthalein indicator: Dissolve 1 gram phenolphthalein in 100 ml of ethyl alcohol and add 100 ml of distilled water.

c- Sodium hydroxide solution (1N): Dissolve 4.0 gram sodium hydroxide to in distilled water to prepare 100 ml of solution.

d- Ammonium molybdate solution: Add 62 ml of concentrated sulphuric acid slowly to 80 ml of distilled water and let cool. Dissolve separately 5 gram of ammonium molybdate in 35 ml of distilled water and mix it with sulphuric acid solution. Add distilled water to make the solution 200 ml.

e- Stannous chloride solution: Dissolve 0.5 gram of stannous chloride in 2 ml of concentrated hydrochloric acid and dilute to 20 ml with distilled water. Use fresh.

f- Standard phosphate solutions: Dissolve 4.388 gram of dried anhydrous potassium hydrogen phosphate in distilled water to make the volume 1 liter. Take 10 ml of this solution and add distilled water to make 1 liter of stock solution containing 1 mg p/l. Prepare phosphorous solutions of various strengths by diluting the stock solution with distilled water, in the range of 0.0 to 1 mg p/liter. Intervals of 0.1 mg p/l.

Method

Take 25 ml of sample in an Erlenmeyer flask and evaporate to dryness. Cool and dissolve the residue in 1 ml of per chloric acid (reagent A).Heat the flask gently so that the contents become colorless. Cool and add 10 ml of distilled water and 2 drops of phenolphthalein indicator (reagent B). Titrate against sodium hydroxide solution (reagent C).until the appearance of slight pink color. Make up the volume to 25 ml by adding distilled water. Add 1 ml of ammonium molybdate solution (reagent D) and 3 drops of stannous chloride solution (reagent E). A blue color will appear. Wait for 10 minutes (never more than 15 minutes) and record absorbance(S) on spectrophotometer at 690 nm. Run simultaneously distilled water blank in similar manner.

Process the standard phosphorus solutions of different strengths (reagent E) in similar manner and plot a standard curve between absorbance and concentrations of standard phosphorus solutions. Deduce the total phosphorus content of sample by comparing its absorbance(S) with standard curve and express the results of total phosphorus in mg/l.

The total particulate phosphorus can be estimated as a difference between the concentration of total phosphorus in unfiltered and filtered samples.

Introduction

Rain water according to its solution equilibrium with the atmospheric air and the absorption coefficient of water for carbon dioxide, contains about 0.6 mg CO_2 per liter. When precipitated water percolates through the soil, additional CO_2 is dissolved out of soil air. Ground waters are, therefore, extra rich in CO_2. Respiratory activity of aquatic organisms and the process of decomposition are important sources of CO_2 in bodies of surface waters. Free CO_2 combines with water partly to form carbonic acids (H_2CO_3), as such in normal practice free CO_2 is distinguished as CO_2 and H_2CO_3.

$$CO_2 + H_2O \rightleftharpoons H_2CO_3$$

Water contains free CO_2 reacts with limestone or chalk ($CACO_3$) of soil or sediment, producing calcium bi-carbonate [Ca $(HCO_3)_2$]. If more CO_2 is present in water than is in equilibrium with air, much more $CaCO_3$ will be dissolved (i.e., converted into calcium bi carbonate) because this extra CO_2 keeps the pH low. As a result, concentration of carbonate (CO_3^{--}) ions will reduce and that of bicarbonate (HCO_3^-) ions will increase. Calcium bicarbonate remains in solution if it is in equilibrium with a certain amount of free CO_2 in water. If the excess CO_2 is removed from the water, the bicarbonate is decomposed into calcium carbonate and free CO_2. Because CO_2 is produced by this process is continues so long as sufficient equilibrium CO_2 present in water.

$$CaCO_3 + H_2O + CO_2 \rightleftharpoons Ca \, (HCO_3)_2$$

    (Weakly soluble)   (Readily soluble)

This reaction will continue until the equilibrium between HCO_3^-, CO_3^{--}, and CO_2 is established.

At varying pH, different proportions of these species of carbon dioxide (free HCO_3^-, CO_3^{--}, and CO_2) present in $CO_2.H_2O$ system are shown in fig. At pH beyond 8.3 because the concentration of free CO_2 is negligible, the bicarbonates begin to decompose and precipitate as carbonates. Between pH 0.0 and 6.35 almost all the species of CO_2 are present in the form of carbonic acid; between pH 6.35 and 10.33 in the form of HCO_3^- ; and between pH 10.33 and 14.00 in the form of CO_3^{--}.

Free CO_2 dissolved in water is the only source of carbon that can be used in photosynthetic activity of aquatic autotrophs. Once fixed by autotrophs it can further be utilized by organisms at other trophic levels. In the absence of free CO_2, the bicarbonates are converted into carbonates releasing CO_2 which is utilized by autotrophs, thus making the water more alkaline.

Requirements:

(a) Laboratory glassware's

(b) Reagents

(1) Sodium hydroxide solution (0.2272 N)- Dissolve 0.909.g of sodium hydroxide in CO_2 free (boiled and cooled) distilled water and make the volume 1 liter. Standardized the solution.

(2) Phenolphthalein indicator- Dissolve 1 g of phenolphthalein in 100 ml of ethyl alcohol and add 100 ml of distilled water. Add NaOH solution (reagent 1) drop by drop until a faint pink color appears.

(3) Method

(4) After collection analyze the sample as soon as possible. Take 50 ml of sample in flask and add 2-3 drops of phenolphthalein indicator (reagent 2). If the color turns pink, free CO_2 is absent in the sample. If the sample remains colorless, titrate it against sodium hydroxide solution (reagent 1) until pink color appears (end point).

Calculation:

$$\text{Free } CO_2 \text{ (mg/liter or ppm)} = \frac{V t \times 1000}{V s}$$

Where, $V t$ = Volume of titrant (ml); and $V s$ = Volume of sample (ml).

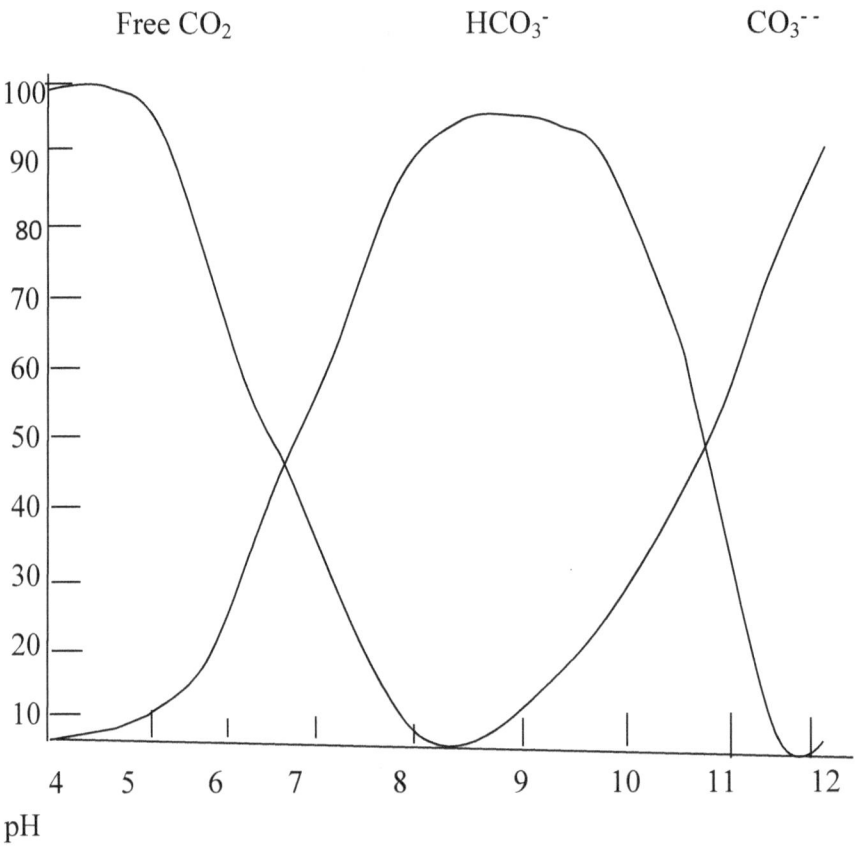

Fig.3 Relationship between pH and percentage of carbon dioxide (free CO_2, HCO_3^-, CO_3^{--}).

Requirements- Reagents:

(a) Potassium permanganate solution (0.125 N) - Dissolve 3.951 g of potassium permanganate (dried at 105° C) in distilled water and make up to 1000 ml. This solution shall be kept in dark and its strength shall be checked periodically.

(b) Standard potassium permanganate solution (0.025 N) - This solution shall be prepared immediately before use by suitable dilution of stock potassium permanganate solution. One ml of this solution is equivalent to 0.1 mg of oxygen.

(c) Dilute sulfuric acid – Add slowly 50 ml of concentrated sulfuric acid to 130 ml of distilled water, cool and make up to 00 ml with distilled water. Add standard potassium permanganate solution until a very faint pink color persists after 4 hours.

(d) Standard sodium thio-sulfate solution (0.0125 N) – Dissolve 3.1g of $NA_2S_2O_3.5H_2O$ in 1 liter of distilled water.

(e) Starch indicator.

Temperature measurement: This determination shall be carried out at a temperature of 37° C.

Procedure: Place 250 ml of the well mixed sample into a clean glass stoppered bottle of 400 ml capacity. Add 10 ml dilute sulfuric acid, followed by an accurately measured volume of standard potassium permanganate solution. Mix by gentle rotation and place in a water bath or in an incubator at 37°C for 4 hours. If the sample contains much suspended matter, it shall be mixed by several times with general rotations during the period of incubation. At the end of 4 hours cool to about room temperature, add a few crystals of potassium iodide and titrate in the bottle with standard sodium thio sulfate solution using a few drops of starch indicator solution. A

63

blank for oxygen absorbed in 3 minutes shall be carried out. Express the result to nearest to 0.05 mg/l.

Calculation: Oxygen absorbed in the 4 hours, mg/l = 0.4 ml of standard potassium permanganate solution consumed in reaction with the sample.

Reactions- $2KMNO_4 + 3H_2SO4 - K_2 SO_4 + 2MNSO_4 + 3H_2O + 5O$

$H_2O + \frac{1}{2} O_2 + 2KI^- - 2KOH + I_2$

$I_2 = O_2$

$2Na_2S_2O_3 + I_2^- - Na_2S_4O_6 + 2NaI$

Chapter- 19   **Dissolved Oxygen (DO)**

Introduction

Dissolved oxygen (DO) is a measure of the amount of oxygen freely available in water. Oxygen dissolved in water is a very important parameter in water analysis as it serves as an indicator of the physical, chemical and biological activities of the water body. The two main sources of dissolve oxygen are diffusion of oxygen from the air and photosynthetic activity.

Diffusion of oxygen from the air into water depends on the solubility of, and is influenced by many other factors like water movement, temperature, salinity and trace elements like zinc etc. Oxygen concentrations are much in air (21%) than in water. Where the air and water meet, this tremendous difference in concentration causes cause's oxygen molecules in air to dissolve into the water. Much of the dissolved oxygen in water comes from the atmosphere due to wind action.

Photosynthesis is carried out in water by autotrophs, depends on the plankton population, light condition, availability of CO_2 etc. Because it requires light, photosynthesis occurs during daylight hours. Oxygen produced during photosynthesis and consumed during respiration and decomposition process, which occur 24 hour a day. This difference bet oxygen production during day and its consumption 24 hour in different processes alone can account large daily variations in dissolved oxygen concentrations. Oxygen production is occurs in the top portions of a water. Where light availability is more. Oxygen consumption is greatest near the bottom of a water body where sunken organic matter accumulates and decomposes.

Dissolved oxygen is extremely useful in self-purification of water bodies. The reduction in dissolved oxygen levels causes anaerobic conditions in water and adversely affects the aquatic inhabitant organisms. The value of dissolved oxygen permitted by WHO are 5-10 mg/l. Levels below 5 mg/l DO not good for fish culture and concentration levels higher than 10 mg/l may also be detrimental to the health of inhabitant organisms because of presence of free radicals.

Determination by the Winkler's method

Requirements:

BOD bottles (100 – 300 ml capacity); laboratory glassware; and reagents as described below:

A- Sodium thio-sulphate solution (0.025 N): Dissolve 6.205 gram of sodium thio-sulphate in previously boiled distilled water and make up the volume to 1 liter. Add a pallet of sodium hydroxide as a preservative. Keep the solution in colored bottle.

B- Manganous sulfate solution: Dissolve 100 gram of manganous sulfate in 200 ml of previously boiled distilled water and filter the solution.

C- Alkaline potassium iodide solution: Dissolve 100 gram of potassium hydroxide and 50 gram of potassium iodide in 200 ml of previously boiled distilled water.

D- Starch indicator: Dissolve 1 gram of starch in 100 ml of warm distilled water and add a few drops of toluene as preservative.

E- Sulfuric acid: Concentrated; special grade 1.84; 18 M.

Method

Take a glass stoppered bottle of 100 ml capacity and fill it with sample avoiding any bubbling. No air should be trapped in bottle after the stopper is placed. Open the bottle and pour in it 1 ml of each manganous sulfate (reagent B) and alkaline potassium iodide (reagent c) solutions by using separate pipettes. A precipitate will appear. Place the stopper and shake the bottle thoroughly. Sample at this stage can be stored for a few days if required. Add 2 ml of sulfuric acid (reagent E) and shake thoroughly to dissolve the precipitate. Transfer the whole content gently in a conical flask (avoid bubbling). Put a few drops of starch indicator (reagent D). Titrate against sodium thiosulfate solution (reagent A) and note the end point when initial blue color turns to colorless.

Calculation:

$$D.O. \ (mg/l) = \frac{V_1 \times N \times 8 \times 1000}{V_2 - V_3}$$

Where, DO= dissolved oxygen; V_1 = volume of titrant (ml); N = normality of titrant (0.025N); V_2 = volume of sampling bottle after placing the stopper (ml); V_3 = volume of manganous sulphate + potassium iodide solutions added (ml).

Introduction

The rate of removal or consumption of oxygen by microorganisms in aerobic degradation of the dissolved or even particulate organic matter in water is called biochemical oxygen demand (B.O.D.). Or biological oxygen demand (BOD) is the measure of organic material contamination in water. Or BOD is the amount of dissolved oxygen required for the biochemical decomposition of organic compounds and the oxidation of certain inorganic materials (e.g., iron and sulfites).

B.O.D. is evaluated by measuring oxygen concentration in the sample before and after incubation in the dark at 20 degree centigrade for 5 days. Preliminary dilution and aeration of sample are usually necessary to ensure that not all the oxygen is consumed during incubation. It is specified in mg/l.

Requirements:

BOD incubator; BOD bottles; laboratory glassware; and reagents as described below:

A- All reagents used in determination of dissolved oxygen.

B- BOD-free water: Pass the deionized glass distilled water through a column of activated carbon and redistill it.

C- Phosphate buffer solution: Dissolve 33.4 gram of $Na_2HPO_4.7H_2O$, 21.5 gram of K_2HPO_4, 8.5 gram of KH_2PO_4 and 1.5 gram of NH_4Cl in BOD free distilled water to prepare 1 liter of the solution. Adjust the pH of the solution to 7.2.

D- Magnesium sulphate solution: Dissolve 82.5 gram of magnesium sulphate in BOD free distilled water to prepare 1 liter of solution.

E- Calcium chloride solution Dissolve 27.5 gram of anhydrous calcium chloride in BOD free distilled water to prepare 1 liter of solution.

F- Ferric chloride solution: Dissolve 0.25gram of ferric chloride in BOD free distilled water.

G- Sulphuric acid (1 N): Add 2.8 ml of sulphuric acid (concentrated) to 100 ml of BOD free distilled water.

H- Sodium hydroxide solution (1 N): Add 4 gram of sodium hydroxide in distilled water and make the volume 100 ml.

I- Allylthiourea solution: Dissolve 500 mg of allylthiourea in distilled water and make the volume 1 liter.

Method

To prepare dilution water aerates the BOD free distilled water in a glass container for about half an hour. Add per liter of this water 1 ml each of phosphate buffer solution (reagent C), magnesium sulphate solution (reagent D), calcium chloride solution (reagent E) and ferric chloride solution (reagent F).

Adjust the pH of sample to neutrality (7.0) using 1 N sulphuric acid (reagent G) or 1N sodium hydroxide solution (reagent H) as the case may be. To ensure that not all the oxygen of the sample is exhausted during incubation, dilute the sample with dilution water at the rate shown in table, according to the expected BOD content of the sample.

Fill two sets of BOD bottles with this sample water and add 1 ml of allylthiourea solution (reagent I) to each bottle. Determine the dissolved oxygen content (D first day) in one set immediately following the Winkler's method of oxygen estimation.

Incubate the other set of BOD bottles at 20 degree centigrade for five days in a BOD incubator. Take out the BOD bottles after 5 days and determine immediately their oxygen content (D5).

Calculation

$$BOD5 \ (mg/l) = (D0\text{-}D5) \times \text{Dilution factor}$$

Where, D0 = initial dissolved oxygen in the sample (mg/l);

and D5 = Dissolved oxygen left out in the sample after 5 days incubation (mg/l).

Introduction

Chemical oxygen demand (COD) is the measure of oxygen required in oxidizing the organic compounds present in water by means of chemical reactions involving oxidizing substances such as potassium dichromate and potassium permanganate. Potassium dichromate is the most suitable oxidant but for waters having more than 2 gram/liter of chlorides potassium permanganate is used, though the results are more variable because latter is self-oxidizing.

The estimation of COD is of great importance for waters having unfavorable conditions for the growth of microorganisms, such as presence of toxic chemicals. In such waters BOD cannot be determined accurately. However, COD too is not a perfect index of organic compounds present in water because, in this reaction, many inorganic compounds are also oxidized, and at the same time some organic compounds remain unaffected. It is specified in mg/l. Both BOD and COD are key indicators of the environmental health of aquatic ecosystems.

Material

A- COD reflux unit consisting of flat bottom flask with ground glass mouth (250 ml) and Liebig (straight tube, single surface) condenser (30 cm).

B- Hot water bath or heating mantle.

C- Reagents as given below:

a- Potassium dichromate solution (0.25N): Dissolve 12.259 gram of potassium dichromate in distilled water to prepare 1 liter of solution.

b- Silver sulfate: Dry, powdered.

c- Mercuric sulfate: Dry, powdered.

d- Sulfuric acid: Concentrated.

e- Ferroin indicator solution: Dissolve 0.695 gram of ferrous sulfate and 1.485 gram of 1, 10-phenonthroline in distilled water to make 100 ml of indicator solution.

f- Ferrous ammonium sulfate solution (0.25 N): Dissolve 98 gram of ferrous ammonium sulfate in distilled water, add 20 ml of sulfuric acid (reagent D), cool and dilute to 1 liter by further adding distilled water.

To standardized this solution, dilute 25 ml of potassium dichromate solution (reagent A) to about 250 ml with distilled water, add 20 ml. of sulfuric acid (reagent D), and cool it. Add 5-6 drops of ferroin indicator solution (reagent E) and titrate against ferrous ammonium sulfate solution (reagent F). The color changes from blue green to a reddish blue at end point.

$$\text{Normality of reagent F} = \frac{VA \times 0.25}{VF}$$

Where, VA= Volume of reagent A ml; and VF= Volume of reagent F used (ml).

Method

Take 20 ml of sample in the flask of reflux unit and add 10 ml of potassium dichromate solution (reagent A), a pinch of each silver sulphate (reagent B) and mercuric sulphate (reagent C) and 30 ml of sulphuric acid (reagent D). Attach Liebig condenser to the mouth of the flask and heat the flask on a hot water bath or heating mantle for at least 2 hours to reflux the contents. Cool the flask, detach from unit and dilute its contents to about 150 ml. by adding distilled water. Add 2-3 drops of ferroin indicator solution (reagent E) and titrate against ferrous ammonium sulphate solution (reagent F). At the end point blue green color of contents changes to reddish blue. Run simultaneously distilled water blank in similar manner.

Calculation

$$COD\ (mg/l) = \frac{(B-A) \times N \times 1000 \times 8}{V.}$$

Where, A= Volume of titrant used against sample (ml); B= Volume of titrant used against blank (ml); N= Normality of titrant (0.25); and V= Volme of sample (ml).

Introduction

This cation occurs in natural waters in far lesser concentration. It behaves in the water as doe's sodium. Though found in small amounts it plays a vital role in the metabolism of freshwater environments and considered to be an important macronutrient.

Potassium is a dietary requirement for us. It is required by RBCs, muscle cells, and brain tissues. Vital function of potassium includes its role in nerve stimulus, muscle contraction, blood pressure regulation, and protein dissolution. In soil and water it acts as bacterial and autotroph nutrient to a certain amount but in excess it causes an increase in eutrophic conditions of ecosystems along with nitrates and phosphates. Skin contact with potassium metals results in caustic potash corrosion. This causes damage to eyes. At high doses potassium chloride interferes with nerve impulses, which interrupts with virtually all bodily functions and mainly affects heart functioning, and also causes hyperkalemia

Determination by Flame Photometer

Principle:

The sample is sucked by an atomizer under controlled conditions. The radiation from the flame enters a dispersing device in order to isolate the desired region of the spectrum. The intensity of the isolated radiation is measured by a phototube. After carefully calibrating the photometer with the solution of known concentration and composition, the concentration of unknown sample can be finding out by comparing the intensity of light produced with that of calibration curve.

Material

A- Equipment's:

Flame photometer; Whatmann filter paper; and reagents as given below:

B- Standard potassium solutions: Dissolve 1.9064 gram of potassium chloride in distilled water to make the volume 1 liter. This stock solution contains 1 gram K/l. Prepare various standard potassium solutions of different strengths by diluting this stock solution with distilled water.

Method

Set the filter of flame photometer for reading at 769 nm and proceed for determination of potassium in sample. Start the compressor and light the burner of flame photometer. Keep the air pressure at 5 lbs. and adjust the gas feeder so as to have a blue sharp flame. Feed the standard potassium solution of the highest value in the range and adjust the flame photometer to read full of emission on scale. Adjust the zero value of the meter by feeding distilled water. Now feed different standard potassium solutions within the ranges 0-1, 0-10, or 0-100 mg K/l one by one and record the emission value for each. Plot a standard curve between concentration and emission of standard potassium solution. Express the result of potassium content in mg/l.

Introduction

Fluoride ions are important in water because of their peculiar characteristics. They cannot be tolerated in too low or too high concentrations. Excess concentration of fluoride (more than 2.00 mg/l) causes dental fluorosis and harm to bony structures (sclerosis). Concentration less than 0.5 mg/l results in dental carries. Hence it is necessary to maintain fluoride concentrations in between 0.5 to 2.00 mg/l.

Determination by Colorimeter

Requirements

I- Colorimeter for use at 570 micro meters.

II- Nesslers tubes capacity: 100 ml.

III- Chemical reagents: As described below-

A- SPANDS SOLUTION: Dissolve 958 mg. SPANDS in 500 ml of distill water.

B- Zirconyl Acid Reagent: Dissolve 133 mg $ZrOCl_2$, 8H2O in 25 ml water. Add 350 ml. Conc. HCl and dilute to 500 ml.

C- Mix equal volume of 1 and 2 to produce a colored reagent. Protect from direct light.

D- Reference solution: Add 10 ml SPANDS solution to 100 ml distilled water. Dilute 7 ml concentrated HCl to 10 ml and add to diluted SPANDS solution.

E- Sodium Arsenite solution: Dissolve 5.0 gram $NaAsO_2$ and dilute to 1000 ml.

F- Stock F- solution: Dissolve 22.00 mg anhydrous NaF and dilute to1000 ml. 1ml = 100 micro grams F-.

G- Standard F- : Dilute stock solution 10 times to obtain 1 ml = 10 micro grams F-.

Procedure

a- Prepare standard curve in the range 0.0 to 1.40 mg by diluting appropriate volume of standard F- in Nesslers tube.

b- Add 10 ml mixed reagents prepared as in C above to all the standard samples, mix well and read density of the bleached color at 570 nm using reference solutions for setting 'zero' absorbance.

c- Plot conc. vs. % transmission or absorption.

d- If the sample contains residual chlorine remove it by adding $NaAsO_2$solution (1 drop or 0.05 ml for each 0.1 mg/l Cl) NaAsO2 concentration should not exceed 1300 mg/l.

e- Take suitable aliquots of sample and dilute to 50 ml in Nessler's tube.

f- Add to 10 ml of the mixed reagent and mix well. Read % transmission or absorption.

g- Calculate the mg/l F- present in the sample using standard curve.

Calculation:

$$F\text{-mg/l} = \frac{\text{mg F- in aliquot} \times 1000}{\text{ml of sample used.}}$$

Introduction

Iron is the second most abundant metal in the earth crust, of which it accounts for about 5%. Elemental iron is rarely found in nature, as the iron ions Fe^{2+} and Fe^{3+} readily combine with oxygen and sulfur containing compounds to form oxides, hydroxides, carbonates and sulfides. Iron is most commonly found in nature in the form of its oxides. It is required by human beings and other organisms for metabolic processes and in the formation of heamoglobin which gives red color to the blood.

Spectrophotometric Analysis

Principle

Iron is brought into reduced form by boiling with acid and hydroxyl amine and treated with 1, 10- phenonthroline at pH 3.2 TO 3.3. There molecules of phenonthroline chelate reacts with each atom of ferrous iron to form an orange red complex. The colored solution obeys Beers law. Its intensity is independent of pH from 3 to 5 and is stable for at least 6 months. A pH between 2.9 and 3.5 ensures rapid color development in the presence of excess phenanthroline.

Requirements:

I- Spectrophotometer (filter 510 nm).

II- Chemical reagents:

A- Concentrated hydrochloric acid.

B- Hydroxyl amine solution- Dissolve 10 gram in 100 ml distilled water.

C- Ammonium acetate buffer solution- Dissolve 250 gram of acetate in 150 ml distilled water and 700 ml glacial acetic acid.

D- Phenanthroline solution- Dissolve 0.12 gram 1:10 phenanthroline monohydrate in 100 ml distilled water and heat to 80 degree centigrade. Heating is not necessary if two drops of concentrated HCl (iron free) are added to distilled water.

E- Stock Iron free solution- Use electric iron wire to prepare the standard. If necessary, clean the wire with sand paper. Weigh 0.2 gram wire and place in 1 l volume flask. Dissolve in 20 ml $6NH_2SO_4$ and dilute to 1 liter. 1 ml=200 micro grams iron.

F- Standard iron solution- These should be prepared the day they used.

Pipette 5.0 ml or 50 ml stock solution into a one liter volumetric flask and dilute to the mark with iron free distilled water. With 50 ml; 1 ml =0.01 mg iron with 5 ml; 1 ml= 0.001 mg.

If the sample is expected to contain less than 2.4 mg of Fe/l, pipette a 50 ml aliquot into 125 Erlenmeyer flask. If the sample is expected to contain a higher concentration of Fe, accurately measure a smaller aliquot containing less than 0.12 gram Fe. Add distilled water to make the volume to 50 ml. Add 2 ml concentrated HCl and few glass beads. Heat to boiling for 5 minutes to bring iron into solution. Cool. Transfer to 100 ml volume flask. Add 1 ml hydroxyl amine hydrochloride, 10 ml ammonium acetate, 10 ml phenonthrolein, dilute to 100 ml, mark. Allow to stand for 5 minutes to permit color development. Compare visually or read on spectrophotometer.

Calculation:

$$\text{Fe mg/l} = \frac{\text{mg. Fe} \times 1000}{\text{ml sample}}$$

All the methods commonly used for the determination of grease are depends upon a preferential solution of the greasy materials using extractions with hexane or petroleum ether (40-60 fractions). Grease generally includes oil, fats, waxes and fatty acids. Industrial waste waters may contain simple esters. The term oil represents a wide variety of substances ranging from low to high molecular weight hydrocarbons of mineral origin that are liquid at ordinary temperatures.

Procedure:

(a) 100 ml of sample is taken to which 5 ml of 1 percent magnesium sulfate is added.

(b) Contents are shaken and mixed thoroughly to which small amount of lime is added so as to form flocks. The precipitate is allowed to settle and is further dissolved in dilute HCl (3 N). Contents are then transferred to the separating funnel, to which are added 50 ml of petroleum ether. Mixture is then shaken vigorously for few minutes and the liquid layers are allowed to separate. The aqueous layer is transferred to another separating funnel and a fresh quantity of 50 ml of petroleum ether added to it. An aqueous layer is separated and the petroleum extracts are combined.

(c) Two grams of powdered anhydrous sodium sulfate is added to it and the contents are shaken intermittently over a period of 30 minutes. Solution is then filtered through Whatmann filter paper No. 1. The filtrate is collected in a flask and is kept on a water bath at about 70°C, so as to evaporate ether. The flask is allowed to cool at room temperature and weighed. The difference in weights (flask with residue less empty flask) is the weight of the residue representing grease.

Calculations:

$$\text{Oil and grease mg/l} = 1000 \ W/V$$

Where, W = Weight in mg of the residue, and V = Volume of the sample in ml taken for the test.

Bibliography

Allen, S.E., Grimshaw, H.M., Parkinson, J.A. and Quarmby, C. (1974) *Chemical analysis of ecological materials.* John Wiley & Sons Inc., N.Y. 565 pp.

APHA, AWWA, WPCF (1995) *Standard Methods for the Examination of Water and Wastewater.* 19th Edn. American Public Health Association American Water Works Association Water pollution Control Federation, Washington DC.

Ara, S., Khan, M.A. and Zargar, M.Y. (2003) *Physico-chemical characteristics of dal lake water. Aqua. Environ. Toxicol,* **12:** 129-134.

Blakely, D.R. and Hrusa, T.C. (1989) *"Inland Aquaculture Development Handbook,"* Fish News Book, 184 Pages.

Bohra, O.P. and Dwivedi, S.N. (1977) *A practical approach for the study of environmental parameters in fresh water ecosystem.* Newsletter No.77/4, Central Inst. Fish. Edu., Bombay. 15 pp.

Boyd, C.E. (1990) *"Water Quality in Ponds for Aquaculture".* Birmingham Publishing Company, Birmingham, Alabama.

Boyd, C.E. (1998) *Water Quality for Pond Aquaculture.* Research and Development Series No. 43. International Center for Aquaculture and Aquatic Environments, Alabama

Agricultural Experiment Station, Auburn University, Alabama.

Carlsberg, S.R. 1972. *New Baltic manual with methods for sampling and analysis of physical, chemical and biological parameters.* Series A, No.29, Co-op. Res. Rep., International Council for the Exploitation of the Sea, Denmark. 145 pp.

Chandrasekhar, J.S., Babu, K.L., and Somashekar, R.K. (2003) *Impact of urbanization on bellandur Lake Bangalore a case study. J. Environ. Biol.,* **24:** 223-227.

Conway, G.R., and Petty, J.N. (1991) *Unwelcome Harvest: Agriculture and Pollution,"* In: *Environmental Protection Agency, Acid Rain, Earth scan,* London, EPA-600/9-79-036, Office of Research and Development, Washington DC.

Cope, O.B. (1966) *Contamination of the freshwater ecosystem by pesticides. J. Appl. Ecol.,* **3:** 33.

Dhindsa, S.S. (1992) "*Fifth Water and Waste Water Analysis Refresher Course*". Public Health Engineering Department Jaipur, Rajasthan, India.

Durve, V.S. and Sharma, L.L. 1976. *Improved under water sampler for limnological work.* Res. And Ind., 21 (2): 93-94.

Garg, S.K. (1998) *"Sewage Disposal and Air Pollution Engineering"*.11th Edn. Khanna Publications, Environmental Engineering., **2:** 188-189.

Golterman, H.L., Clymo, R.S. and Ohnstad, M.A.M. 1978. *Methods for physical and chemical analysis of fresh waters*, 2nd ed. Blackwell Scientific Publications, Oxford. 213 pp.

Holdgate, M.W. (1979) *"A Perspective of Environmental Pollution,"* Cambridge University Press, Cambridge.

Neera, S., Meena A. and Anupama, T. (2003) *Study of physico chemical characteristics of water bodies around Jaipur. J.Environ.Biol.*, **24:**177-180.

Neeri (1979) *Course Manual Water and wastewater analysis. Neeri,* Nagpur, pp: 134.

Pillay, T.V.R. (1992) *"Aquaculture and the Environment,"* Type Sector Limited, Hong Kong, **67 Pages**.

Russel E. Train (1979) *"Quality Criteria for Water"*. U.S. Environmental Protection Agency, Washington D.C. Castle House Publications Ltd., Great Britain.

Sandra, E. Shumway (2007) *"A review of the effects of algal blooms on shellfish and aquaculture"*. *Journal of the world aquaculture society,* **21:**65-104.

Saxena M.M. (1990) *"Environmental Analysis Water, Soil, and Air"*. Agro Botanical Publishers, India.

Saxena, S. (1998) *Settling studies on pulp and paper mill wastewaters. Indian J. Environ. Health.* **20:**273-280.

Verma, R. (1978) *Physico chemical and biological characteristics of Kedarabad drain. Op. Cit.,* **20:** 1-13.

WHO (1993) *Guidelines for drinking water quality.* Water Sanitation Health. **Vol. 1**.